Name: Eng. / Mostafa Yacoub
Abdellatif Mahmoud

I0427008

Nationality: Egyptian

ORCID: 0000-0002-9991-4624

Email:

moshhaabma2015@gmail.com

Qualification: civil engineer Cairo
University 2003

- <u>**Bounded and large gaps between prime numbers**</u>

- In this paper or research, we will discuss the value of the bounded and large gaps between prime numbers.

- We will do the work based on my discovered formula that connects between prime and composite numbers

My discovered formula:

Definitions:

Array PTBP

It is the following Array of odd numbers

$$\begin{vmatrix} 1 & 3 & 7 & 9 \\ 11 & 13 & 17 & 19 \\ 21 & 23 & 27 & 29 \\ 31 & 33 & 37 & 39 \\ 41 & 43 & 47 & 49 \\ 51 & 53 & 57 & 59 \end{vmatrix}$$

And so on....

- **For a given set of consecutive primes whose numbers =n that start with prime 3 and end with prime F and not including prime 2 and prime 5**

i.e.

set=[3,7,11,13,...

........................,F]

S=product of those consecutive primes

i.e

$$S = \prod_{i=3}^{i=F} (i)$$

Range=R_k = 10 × S × k

Where k = [1, 2, 3, 4, ..., ∞(infinity)

i.e R_1=10 x S x 1 and R_2=10 x S x 2

And so on

- **Number of composite numbers that belong to Array PTBP and created by the effect of those consecutive primes within the range R_K**

$$= \left[\left(K \times 4^{\times \frac{S}{3}} \right) + \left(\sum_{j=7}^{j=F} \left(K \times 4 \times \left(\frac{S}{j} \right) \times \prod_{i=7}^{i=\text{prime number befor current prime number } j} \left(\frac{i-1}{i} \right) \right) \right] - (n)$$

Where j =consecutive values of primes
7, 11, 13,..............., F
And i= consecutive values of primes
3, 7, 11, 13,........, prime before current j prime

- The previous formula can be applied for any number of consecutive prime numbers that start with prime number 3

- The first term $(k \times 4 \times \frac{S}{3})$ represents the count of unique Composite numbers +1 that belong to the Array PTBP and are created by prime number 3 within the range

$$R_k = 10 \times S \times k$$

- The second term

$$\sum_{j=7}^{j=F} (K \times 4 \times (\frac{S}{j}) \times$$

$i = prime\ number\ befor\ current\ prime\ number\ j$

$$\prod_{i=7} \qquad (\frac{i-1}{i})$$

Represent the count of unique Composite numbers+n-1 that belong to the Array PTBP and are created by each prime number after the prime number 3 within the range

$R_k = 10 \times S \times k$

- **The third term (-n)**
  Subtracting n (number of consecutive primes starting from prime number 3) because the count of composite numbers generated from those consecutive primes includes the count of those primes in the range
  $$R_k = 10 \times S \times k$$
- **Explanation and proof for my theory in my previous paper (prime number theory)**

- **We will mention only the concept of number cycle**

  We can use the number cycle concept to understand the behavior of consecutive primes in creating composite numbers.

  i.e.

  $$S = \prod_{i=3}^{i=F} (i)$$

  Range=cycle range= R_k = 10 × S × k

  Where k= [1, 2, 3, 4, ..., ∞(infinity)

  i.e. R_1=10 x S x 1 and R_2=10 x S x 2

  And so on

- **Now consider only one k value =1**

For a set of consecutive primes and according to my formula the result will be

- $$=[(K \times 4^{\times \frac{S}{3}}) + ($$

$$\sum_{j=7}^{j=F} (K \times 4 \times (\frac{S}{j}) \times$$

$i = prime\ number\ befor\ current\ prime\ number\ j$

$$\prod_{i=7} \qquad (\frac{i-1}{i})$$

$$)]-(n)$$

And

including the count of prime numbers within the set

$$=[(K \times 4^{\times \frac{S}{3}}) + ($$

$$\sum_{j=7}^{j=F} (K \times 4 \times (\frac{S}{j}) \times$$

$i = prime\ number\ befor\ current\ prime\ number\ j$

$$\prod_{i=7} \qquad (\frac{i-1}{i})$$

)]

Which represent the count of numbers (that belong to the Array PTBP) that is divisible of the prime numbers that belong to the set of consecutive primes

- **That count repeated each following cycle (k=2,3, ... ∞(infinity) and represents the count of numbers that belong to the array PTBP and are divisible by the prime numbers within the set of consecutive primes**

- $t$he count of the complementary part that repeated each cycle (k=1,2,3, ... ∞(infinity)

$$= (4 \times \prod_{i=3}^{i=F} (i-1))$$ as explained in my paper (infinite primes)

Which represent the count of numbers (that belong to the Array PTBP) that is not divisible by each prime number that belong to the set of consecutive primes

- And now if we take the following set of consecutive primes

Set_1=[3,7,11,13,.............................
.........................,F]

$$S_1 = \prod_{i=3}^{i=F} (i)$$

Range =

$R_k = 10 \times S_1 \times k$

Where

k=[1,2,3,4, ..., ∞(infinity)

And let A = the count of odd numbers that belong to array PTBP within each cycle (i.e k=1,2,3,.....

∞(infinity) and not divisible of each prime number that belongs to set_1 i.e

$$A= (4 \times \prod_{i=3}^{i=F} (i-1))$$

and let P = the next consecutive prime after prime then we have set_2

set_2=

[3,7,11,13,.....................................

.........,F,P]

$$S_2 = \prod_{i=3}^{i=P} (i)$$

Range=Rk = $10 \times S_2 \times k$

Where

k=[1,2,3,4, ..., ∞(infinity)

- and let B = the count of odd numbers that belong to the array PTBP and are not divisible of each prime number that belongs to set_1 within each cycle created by consecutive primes that belong to set_2 = A × P

  i.e

  B = A $\times P =$

$$(4 \times P \times \prod_{i=3}^{i=F} (i-1))$$

And we have The count of composite numbers created by the prime number P within each cycle after the first cycle (because the first cycle contains

that particular prime number P)=

C=

$$C = (4 \times S_2 / P) \times \prod_{i=3}^{i=F} \left(\frac{i-1}{i}\right)$$

=

$$(4 \quad \times P \times \prod_{i=3}^{i=F} (i) \quad / \quad P) \quad \times$$

$$\prod_{i=3}^{i=F} \left(\frac{i-1}{i}\right) = 4 \quad \times \prod_{i=3}^{i=F} (i-1)$$

Then B-C=

$$(4 \times \prod_{i=3}^{i=F} (i-1)\ ) \times (P\text{-}1) = (4 \times \prod_{i=3}^{i=P} (i-1)\ )$$

So prime number P must exist within the first half part within the first cycle created by the set of consecutive primes set_1 due to the symmetry of the distribution of that count B-C

- i.e

  P-F=Min can be 2 or 4 as the minimum value as a probability

And the maximum possible value for prime P can be estimated as follows, also as a probability

let N= $\left(4 \times \displaystyle\prod_{i=3}^{i=F} (i-1)\right)$ / 2

=$\left(2 \times \displaystyle\prod_{i=3}^{i=F} (i-1)\right)$

N must be divisible by 4

$$Max = \left(\left(10 \times \prod_{i=3}^{i=F} (i)\right)/2\right)$$

$$-((N/4) \times 10)$$

$$Max = \left(\left(10 \times \prod_{i=3}^{i=F} (i)\right)/2\right)$$

$$-\left(\left(5 \times \prod_{i=3}^{i=F} (i-1)\right)\right)$$

$$Max = 5 \times \left[\left(\prod_{i=3}^{i=F} (i)\right) - \left(\prod_{i=3}^{i=F} (i-1)\right)\right]$$

- **This estimated value can be reduced by studying the**

distribution of the numbers that belong the the cycle produced by the consecutive primes that belongto

Set_1=[3,7,11,13,...
............, F]

And are not divisible of each of those consecutive primes

Then we approximate the Max value to the next bigger odd number that belongs to the array PTBP = Z

- so the maximum possible value for prime number P = Z= Max + 2
- and the minimum possible value = F + 2 or =F + 4 depending on the last digit of prime number F

- and the max possible gap between prime number F and prime number P

$$\text{max gap} = Z - F = \text{Max} + 2 - F$$

- And now we will use Bertrand's postulate
  To prove that the actual gap is much smaller than
  Max gap
- According to Bertrand's postulate
  Next prime P
  And modified the equation to satisfy the fact that the last digit of prime numbers that belong to

the array PTBP must be 1 3 or 7 or 9

$P \leq (2 \times F) - 1$ if the last digit of $F = 1$ or 7 or 9

And

$P \leq (2 \times F) - 3$ if the last digit of $F = 3$

And the ratio between the two values (max possible gap and actual)

$$= ((2 \times F) - 3) / \left[\left(5 \times \left[\left(\prod_{i=3}^{i=F} (i) \right) - \prod_{i=3}^{i=F} (i-1) \right] \right) + 2 \right]$$

if the last digit of F = 3

and =

$$=(\ (2 \times F) -1)/[(\quad 5 \times [(\ $$

$$\prod_{i=3}^{i=F} (i)$$

$$)-$$

$$\prod_{(i=3}^{i=F} (i-1))$$

$$])\ +2])$$

if last digit of F = 1 or 7 or 9

- **and it is clear that as we take more bigger set of consecutive primes the actual max possible gap between prime number P and prime number F is relatively very small**

- **for example**

  If F =13

  then

  The ratio $\times 100$ **=100** $\times$ **(23)/((5** $\times$ **((3** $\times$ **7** $\times$ **11** $\times$ **13)** $-$ **(2** $\times 6 \times 10 \times 12$)) **+2)= 0.294 %**

And if F=19

then

The ratio $\times 100 = 100 \times (37)/((5 \times ((3 \times 7 \times 11 \times 13 \times 17 \times 19) - (2 \times 6 \times 10 \times 12 \times 16 \times 18)))+2)$

= 0.0013 %

And if F=43

The ratio = 1.97E-12 %

- **This means that the gap between prime numbers**

becomes relatively smaller as the value of the prime numbers increases.

www.ingramcontent.com/pod-product-compliance
Lightning Source LLC
Chambersburg PA
CBHW071020290526
45795CB00005B/1879